The Little Fun Book of Molecules/Humans: Molecules/Humans

by

John Hodgson

This is a book of fiction. Places, events, and situations in the story are purely fictional. Any resemblances to actual persons living or dead, is coincidental.

C 2002 by John Hodgson.

All rights reserved.

ISBN: 978-1-387-38331-3

PREFACE

This book is intended to look at the similarities existing between two entities. If we understand more of each entity there may be clues to hold new clues to hold new scientific information leading to new research.

The Fun Little Book of Molecules/Humans

Molecules/humans are prone to disease. Molecules/humans cannot fight off some diseases without outside help.

The Little Fun Book of Molecules/Humans

Molecules/humans don't always get along. That's why they don't come together.

The Little Book of Molecules/Humans

Different molecules/humans behave differently having different reaction/behavior to changing chemicals/situations.

The Fun Little Book of Molecules/Humans

In the world molecules/humans are different/equal compared to each other.

The Fun Little Book of Molecules/Humans

In communities, molecules/humans differ/gather with each other joining/making friends.

The Little Fun Book of Molecules/Humans

When molecules/humans play they move quickly/play basketball doing what is needed to form cells/shoot a basket.

The Fun Little Book of Molecules/Humans

Molecules/humans form substances/make things to form life/develop.

The Little Fun Book of Molecules/Humans

When molecules/humans need/stretch the imagination they form/develop.

The Little Fun Book of Molecules/Humans

When molecules/humans mesh they have chemical or physical reaction/produce.

The Little Fun Book of Molecules/Humans

Molecules/Humans share from one another taking what they need from each other then separating.

The Little Fun Book of Molecules/Humans

Good molecules/humans excel and teach one another.
Bad molecules/humans don't interact and fall down.

The Little Fun Book of Molecules/Humans

Molecules/humans are different sizes and shapes.
Molecules/humans attract one another.

The Little Fun Book of Molecules/Humans

There are many molecules/humans on earth. Each with it's own distinction. Some molecules/humans are bad/good.

The Little Fun Book of Molecules/Humans

Some molecules/humans produce offspring. Some more than others, some doubles and triplets.

The Fun Little Book of Molecules/Humans

When molecules/humans breathe they accept oxygen. Turning the oxygen to useful energy, then living long and prosperous.

The Little Fun Book of Molecules/Humans

Molecules/humans need each other to form new substances/get married.

The Little Fun Book of Molecules/Humans

Molecules/humans group together/play hockey to make solvents/have fun.

The Little Fun Book of Molecules/Humans

When molecules/humans come in contact some get along/bond and form substances/relationships to last a lifetime.

The Little Fun Book of Molecules/Humans

With experiments we can understand these molecules/humans like we never knew before.

The Little Fun Book of Molecules/Humans

Why do molecules/humans behave the way they do?

The Little Fun Book of Molecules/humans

Molecules/humans sometimes devour/ step on each other to get what they want.

The Little Fun Book of Molecules/humans

Molecules/humans get hyperactive when fed certain proteins/sugar.

The Little Fun Book of Molecules/Humans

Molecules/humans pair together to create different chemicals/babies and fall in love.

The Little Fun Book of Molecules/Humans

Molecules/humans have cancer/war destroying many molecules/ humans.

The Little Book of Molecules/Humans

Some Molecules/humans become cells/friends while other molecules/humans do not like each other/mixtures.

The Little Fun Book of Molecules/Humans

Some molecules/humans transfix on each other when in contact/speak.

The Little Fun Book of Molecules/ Humans

We must find out about Molecules/humans so we can understand disease/cures.

The Little Fun Book of Molecules/Humans

The molecules/humans/cells live on earth forming new complex molecules/offspring.

The Little Fun Book of Molecules/humans

Some molecules/humans are more successful in the world/have more blessings.

The Little Fun Book of molecules/Humans

Molecules/humans are combined with each other forming animals/nations.

The Little Fun Book of Molecules/Humans

Complex molecules/humans know plenty of the same elements/creatures.

The Little Fun Book of Molecules/Humans

There are molecules/humans that go to experiments/school to under stand each molecule/human.

The Little Fun Book of Molecules/Humans

Each element/human needs a nucleus/teacher to live, develop and learn.

The Little Fun Book of Molecules/Humans

An element/pupil gives a nucleus/teacher information that is needed for development/body.

The Little Fun Book of Molecules/Humans

Without the molecules/humans living/learning there would be no life/future.

The Little Fun Book of Molecules/Humans

Us Molecules/humans must understand the similarities between each other.

The Little Fun Book of Molecules/Humans

Molecules/humans that pair together have love/positive relationships.

The Little Fun Book of Molecules/Humans

Molecules/humans that enjoy each other's company form solutions/enjoy things together.

The Little Fun Book of Molecules/humans

Molecules/humans have needs to bond/love.

The Little Fun Book of Molecules/Humans

Molecules/humans needs are amazing/tremendous.

The Little Fun Book of Molecules/Humans

To understand molecules/humans we must do experiments/learn from each other.

The Little Fun Book of molecules/humans

Some complex molecules/humans make special relationships/get married.

The Little Fun Book of Molecules/Humans

Some molecules/humans never find a special relationship/remain on their own.

The Little Fun Boob of Molecules/Humans

Some molecules/humans never get to bond/get what they want.

The Little Fun Book of Molecules/Humans

Some molecules/humans show their true colors/get what they need.

The Little Fun Book of Molecules/Humans

Molecules/humans need each other/get lonely.

The Little Fun Book of Molecules/Humans

Molecules/humans lay dormant/sit down.

The Little Fun Book of Molecules/Humans

Molecules/humans can get along/do not always see together eye to eye.

The Little Fun Book of Molecules/Humans

When molecules/humans do not think alike/might separate.

The Little Fun Book of Molecules/Humans

Some molecules/humans like each other and bond/leave each other bitter.

The Little Fun Book of Molecules/Humans

Some molecules/humans travel with others/travel alone.

The Little Fun Book of Molecules/Humans

Some molecules/humans join together/liking each others company.

The Little Fun Book of Molecules/Humans

Molecules/humans join with each other form solutions/communicate.

The Little Fun Book of Molecules/Humans

Molecules/humans like to bond with things they like/do different things.

The Little Fun Book of Molecules/Humans

Molecules/humans push each other around/bully one another.

The Little Fun Book of Molecules/Humans

Some molecules/humans get what they want/ go after what they want.

The Little Fun Book of Molecules/Humans

Molecules/humans take in elements/food.

The Little Fun Book of Molecules/Humans

Molecules/humans engage in different behavior/sex.

The Little Fun Book of Molecules/Humans

Molecules/humans develop to be strong/men.

The Little Fun Book of Molecules/Humans

Molecules/humans forge innovation/have specific goals.

The Little Fun Book of Molecules/Humans

Molecules/humans enrich one another/changing with time.

The Little Fun Book of Molecules/Humans

Molecules/humans employ each other/have standards.

The Little Fun Book of Molecules/Humans

Molecules/humans have special needs/food/water.

The Little Fun Book of Molecules/Humans

Molecules/humans make change/common bonds.

The Little Fun Book of Molecules/Humans

If molecules/humans try hard to get what they need, molecules and humans will find that there is beauty/peace.

The Little Fun Book of Molecules/Humans

If molecules/humans can understand what they need from each other and not what they want from each other there would be a symbiotic relationship/nurturing relationship.

The Little Fun Book of Molecules/Humans

Molecules/humans move around to different molecules/humans/migrate.

The Little Fun Book of Molecules/Humans

Living molecules/humans need nutrients/food.

The Little Fun Book of Molecules/Humans

Molecules/humans form chains/dance.

The Little Fun Book of Molecules/Humans

Molecules/humans are complicated/sophisticated.

The Little Fun Book of Molecules/Humans

Molecules/humans need energy/work.

The Little Fun Book of Molecules/Humans

Without nutrients, molecules/humans will die/suffer.

The Little Book of Molecules/Humans

Molecules/humans do travel/move/migrate to one another.

The Little Fun Book of Molecules/Humans

Molecules/humans feed off one another having relationships/parenting.

The Little Fun Book of Molecules/Humans

Molecules/humans live from one another throughout their lifetime.

The Little Fun Book of Molecules/Humans

Each Molecules/human is like nucleus/earth living with each other. So humans must understand the earth we all live on.

The Little Fun Book of Molecules/Humans

Molecules/humans like the goodness that is felt from each other/get married.

The Little Fun Book of Molecules/Humans

Molecules/humans have memories/sometimes ignore each other.

The Little Fun Book of Molecules/Humans

Molecules/humans know what they need or get what they need/find peace to be wonderful.

The Little Fun Book of Molecules/Humans

Molecules/humans forming complexities/going to concert.

The Little Fun Book of Molecules/Humans

Molecules/humans agree with each other forming bonds/friends.

The Little Fun Book of Molecules/Humans

Certain molecules/humans attract one another/elements/marriage.

The Little Fun Book of Molecules/Humans

Molecules/humans act much alike given the right elements.

The Little Fun Book of Molecules/Humans

Molecules/humans fall into many categories/occupations.

The Little Fun Book of Molecules/Humans

Molecules/humans exist to better life/live longer.

The Little Fun Book of Molecules/Humans

Molecules/humans are living to find similarities/experiments.

The Little Fun Book of Molecules/Humans

Molecules/humans hold answers to benefit all/learn.

The Little Fun Book of molecules/Humans

Molecules/humans shouldn't give up/hope.

The Little Fun Book of Molecules/Humans

Molecules/humans have power/with speech/words.

The Little Fun Book of Molecules/Humans

Molecules/humans should participate/live.

The Little Fun Book of Molecules/Humans

Molecules/humans are forever evolving/finding new theories.

The Little Fun Book of Molecules/Humans

Molecules/humans are changing/evolve.

The Little Fun Book of Molecules/Humans

Molecules/humans govern life/harness life.

The Little Fun Book of Molecules/Humans

Molecules/humans are in development/evolving.

The Little Fun Book of Molecules/Humans

Molecules/humans get frustrated/finding cures.

The Little Fun Book of Molecules/Humans

Molecules/humans hold answers to questions/experiments.

The Little Fun Book of Molecules/Humans

Molecules/humans have talents/GOD given.

THE END

Lightning Source UK Ltd.
Milton Keynes UK
UKHW020700090223
416722UK00005B/520